Science Matters

NORTHERN LIGHTS

David Whitfield

WEIGL PUBLISHERS INC.

Published by Weigl Publishers Inc.
350 5th Avenue, Suite 3304, PMB 6G
New York, NY USA 10118-0069
Website: www.weigl.com

Library of Congress Cataloging-in-Publication Data

Whitfield, David.
  Northern lights / by Dave Whitfield.
       p. cm. -- (Science matters)
  Includes bibliographical references and index.
  ISBN 1-59036-413-9 (alk. paper) -- ISBN 1-59036-419-8 (pbk. : alk. paper)
  1. Auroras--Juvenile literature. I. Title. II. Series.
  QC971.4.W55 2007
  551.56'7--dc22

                              2005029920

Printed in China
1 2 3 4 5 6 7 8 9 10 09 08 07 06

Editor Frances Purslow
Design and Layout Terry Paulhus

Cover: Northern lights in Majestic Valley, Alaska

Photograph Credits
University of Heidelberg, ESA: page 12T; **NASA,ESA, M. Robberto (Space Telescope Science Institute/ESA) and the Hubble Space Telescope Orion Treasury Project Team plus C.R. O'Dell (Rice University), and NASA**: pages 12 & 13 background; **NASA, ESA, J. Hester and A. Loll (Arizona State University)**: page 14L.; **Jacob Bourjaily**: page 19.

All of the Internet URLs given in the book were valid at the time of publication. However, due to the dynamic nature of the Internet, some addresses may have changed, or sites may have ceased to exist since publication. While the author and publisher regret any inconvenience this may cause readers, no responsibility for any such changes can be accepted by either the author or the publisher.

Every reasonable effort has been made to trace ownership and to obtain permission to reprint copyright material. The publishers would be pleased to have any errors or omissions brought to their attention so that they may be corrected in subsequent printings.

Contents

Studying Northern Lights

High up in a cold, dark sky, nature puts on a dazzling show of shifting light. Green lights appear. They move around in the sky and are joined by flashes of red, pink or purple. These are the northern lights. They are also called aurora borealis. Northern lights are usually seen during the winter months. They occur in northern **polar regions** on Earth.

■ Every display of northern lights is unique. The glowing colors change shape, fade, and then brighten again.

Northern Lights Facts

Did you know that some people claim they have heard sounds from northern lights? Some describe it as a swishing sound. Others say it is a crackling sound. Here are more interesting facts about northern lights.

- Northern lights can be seen from space. Scientists use instruments in spacecraft to study them.

- Long ago, some Arctic people thought northern lights were spirits playing soccer in the sky.

- Scientists can predict when northern lights will appear. Information from the Sun and the **solar wind** in space is used.

- Few people ever see the southern lights, or aurora australis. Most of the time, these lights appear over areas where no people live.

How Northern Lights Form

Nature creates northern lights in the **atmosphere** high above Earth. The atmosphere contains many gases. The main gases in the atmosphere are oxygen and nitrogen. Northern lights occur when energy-charged **particles** from space strike the gases. The colliding particles cause the gases to glow in different colors.

■ Glowing northern lights can stretch for 1,000 miles (1,600 kilometers) across the sky.

Northern Lights Myths

Northern lights seemed magical to people long ago. They did not know what the lights were, so they created **myths** and legends about them. People in different parts of the world had their own unique stories about the beautiful lights in the sky.

In Asia, some people thought northern lights were caused by dragons fighting in the sky.

A legend belonging to the Menominee Indians in the United States tells that northern lights were giant torches used by friendly giants to help spear fish at night.

In Finland, northern lights are called "fox fires." People thought the lights were made by an Arctic fox starting fires or scattering snow with its bushy tail.

The Sun

Northern lights begin on the Sun. The Sun is the star closest to Earth. It is 93 million miles (150 million km) from Earth. The Sun controls Earth's weather and **climate**. It is so big Earth could fit inside it one million times. The Sun is a huge ball of fiery gases.

On the Sun's surface are storms called sunspots. During these storms, explosions called solar flares shoot energy-charged particles far out into space. It is these particles that create northern lights when they enter Earth's atmosphere.

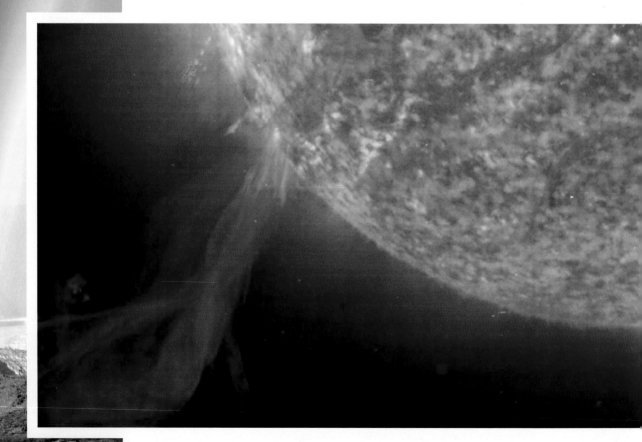

● Astronomers at NASA's Solar & Heliospheric Observatory study solar flares.

Solar Wind

Particles shot into space from the Sun are carried away by the solar wind. This wind travels 200 to 500 miles (320 to 800 km) per second.

Billions of particles are carried in the solar wind. Some of them travel to Earth. It takes two to three days for the particles to travel through space to Earth.

Some of these particles also travel to other planets. In fact, auroras occur on all of the planets in our **solar system**.

The NASA spacecraft *Mariner 2* first tracked these particles coming from the Sun in 1962. It measured the solar wind while on a flight to Venus that lasted three and a half months.

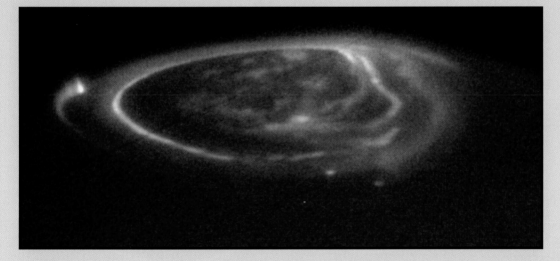

■ Auroras often form over Jupiter's north pole.

Magnetic Fields

Earth is like a giant magnet with ends or poles that pu toward each other. This attraction forms an invisible **magnetic field** around Earth. The field is made up of magnetic lines that run between the poles. It is strongest near the north and south magnetic poles.

Some of the energy-charged particles in the solar wind are trapped by Earth's magnetic field. They are then drawn toward the poles.

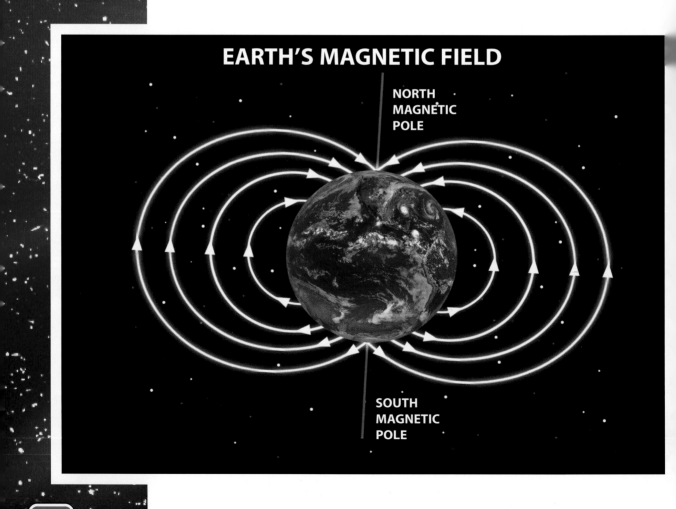

EARTH'S MAGNETIC FIELD

NORTH MAGNETIC POLE

SOUTH MAGNETIC POLE

The Auroral Oval

As particles from the Sun enter Earth's atmosphere, they are guided into beams by the planet's magnetic field. The particles gather around the north and south magnetic poles in the shape of an oval.

In these ovals, particles strike the gases, causing them to glow as northern lights. Auroral ovals are 1,000 to 2,000 miles (1,600 to 3,200 km) above the poles. From high above Earth, northern lights look like a giant glowing ring around the top of the planet. There is a similar auroral oval over the magnetic south pole.

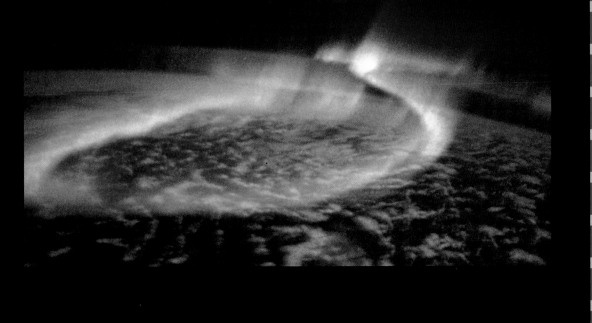

Sky Technology

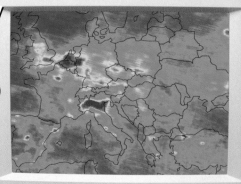

Geographic Information System (GIS)
Special computers called Geographic Information Systems (GIS) gather information about Earth. Scientists use GIS to map **air pollution** in cities and towns. Results are posted on the Internet so people can read about the types and amount of pollution where they live.

Telescopes
Telescopes help us see objects that are far away. Astronomers use them to observe space objects, such as stars, planets, and whole **galaxies**. Telescopes make distant objects appear closer by collecting light. Telescopes can collect more light than the human eye can.

Weather Satellite

Weather satellites are spacecraft that circle Earth. They provide a weather watch on the entire planet. Weather satellites take photographs of Earth's atmosphere. These help meteorologists predict storms and other weather patterns. These satellites also carry special instruments that record information on computers. They monitor events in the atmosphere, such as auroras, dust storms, pollution, and cloud systems.

Radar

Meteorologists gather huge amounts of information in order to predict the weather. **Radar** can tell them what is inside a cloud. This can be rain or hail. Radar can also track a storm that is coming. It helps meteorologists warn people if the storm is dangerous.

Gases in the Atmosphere

Earth's atmosphere is made up of 78 percent nitrogen, 21 percent oxygen, and 1 percent of other gases, such as hydrogen and helium.

As the particles riding the solar wind enter Earth's atmosphere, they collide with these gases. The collision cause the particles to release energy. This energy causes the gases to glow. The glowing lights are called auroras.

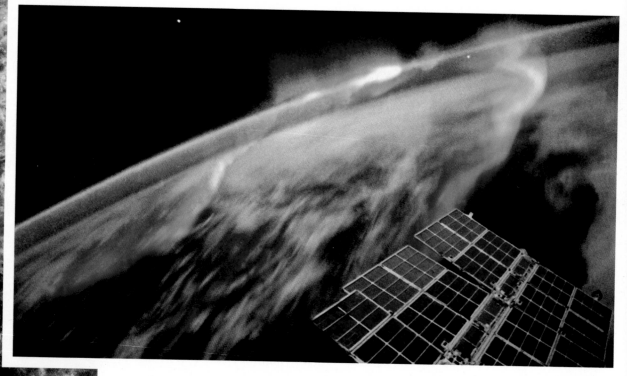

■ The glowing lights of the aurora australis over Australia can be seen from spacecraft in orbit above Earth.

Aurora Colors

Most northern lights are shades of green. However, the color of northern lights depends on which gas is glowing. It also depends on how high above Earth the gas is located. Some gases, such as oxygen, occur at various **altitudes**. Oxygen below 150 miles (240 km) glows yellowish-green. Oxygen above this height glows red.

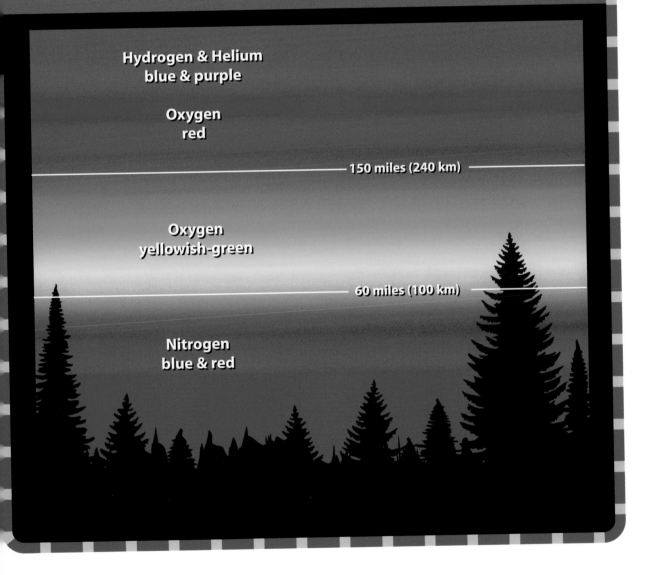

Hydrogen & Helium
blue & purple

Oxygen
red

150 miles (240 km)

Oxygen
yellowish-green

60 miles (100 km)

Nitrogen
blue & red

Viewing Auroras

People who live in the far north have the best view of the northern lights. Auroras are often in the night sky in this part of the world. During a period of large sunspots and solar flares, people living farther south may also see northern lights.

Fall, winter, and early spring are the seasons to watch for northern lights. The best time to see the lights is near midnight, on a night when there is no moon and the sky is clear. However, northern lights can occur at any time of night.

■ City lights make it more difficult to see auroras. Viewing northern lights outside the city is easier because it is darker.

Types of Northern Lights

Northern lights are not all the same. They occur in different shapes.

Auroral arcs look like a shining curtain across the sky.

A **corona** looks like an explosion of light, with rays going in all directions from a central point.

Curved arcs can be hundreds of miles long. They look like trails through the sky.

When arcs are broken up into lines, they are known as **rayed bands**. These are often brighter and more colorful than arcs.

Predicting Northern Lights

Particles from the Sun carried by the solar wind are the main energy source for northern lights. Scientists study the sunspots and solar flares that produce solar wind, in order to predict when northern lights will happen. When the solar wind is calm, there are few northern lights. When the wind is strong, Earth can have intense northern lights.

As solar winds approach Earth, they pass a satellite. This satellite "reads" the winds and alerts scientists when auroras are likely to occur.

■ The *Spartan 201-05* satellite leaves the cargo bay of the space shuttle *Discovery*. It records data on solar wind.

A Life of Science

Kristian Birkeland

Professor Kristian Birkeland was a Norwegian scientist and explorer. In 1896, he was the first person to create auroras in a laboratory. He was testing his **theory** about northern lights and solar wind.

Birkeland predicted that northern lights were caused by particles from the Sun being captured in Earth's magnetic field. These particles moved along the magnetic field into regions around the north and south poles. Kristian Birkeland built a model to test his theory.

Birkeland's image was added to the 200 Norwegian Kroner in 1994 in recognition of his achievements.

Surfing Northern Lights

How can I find more information about northern lights?

- Libraries have many interesting books about northern lights.
- Science centers and museums are great places to learn about northern lights.
- The Internet offers some great websites dedicated to northern lights.

Where can I find a good reference website to learn more about northern lights?

Encarta Homepage
www.encarta.com

- Type "northern lights" or "auroras" into the search engine.

How can I find out more about northern lights?

- Poker Flat Research Range of the University of Alaska
  www.pfrr.alaska.edu/main.htm
  Click on Aurora Information, then on Aurora Forecast.

Science in Action

Paint Northern Lights

Use watercolors to paint a scene of northern lights.

You will need:
- paintbrush
- water
- a piece of white art paper
- red, blue, and green paint
- scissors
- a piece of black construction paper
- glue

Brush plain water on the white paper to wet it completely. While the paper is still wet, brush red, green, and blue paint in shapes that look like northern lights. The colors will run and blend together on the wet paper.

Allow the paper to dry.

Cut the black paper to make a border or frame around the northern lights. Glue it in place on the painted paper. Cut out a house or trees to use as silhouettes against the northern lights background. Glue them in place.

What Have You Learned?

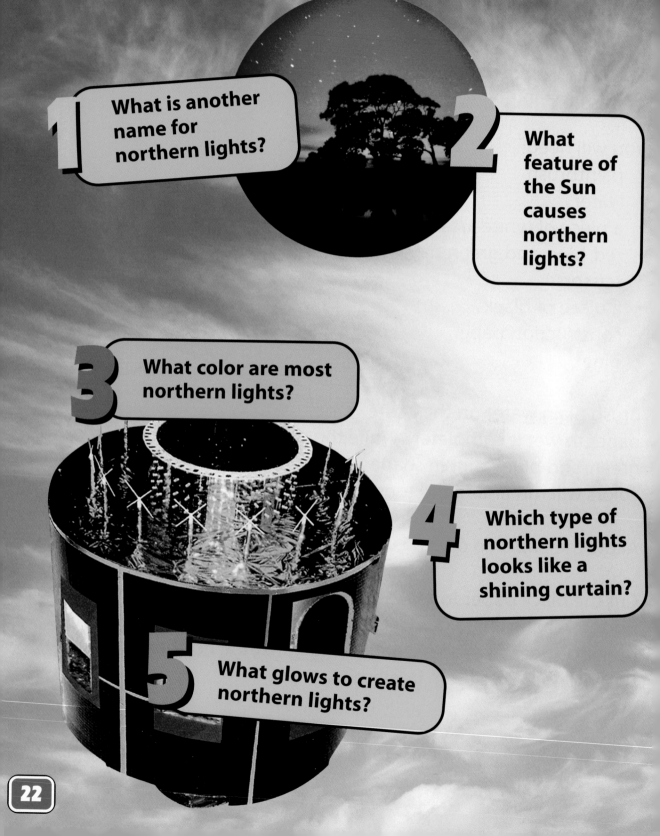

1 What is another name for northern lights?

2 What feature of the Sun causes northern lights?

3 What color are most northern lights?

4 Which type of northern lights looks like a shining curtain?

5 What glows to create northern lights?

6 What was Kristian Birkeland's theory about northern lights?

7 Is Earth the only planet with auroras?

8 Where do northern lights occur?

9 Which season is best for viewing northern lights?

10 How do scientists predict when there will be northern lights?

Words to Know

air pollution: harmful materials, such as chemicals and gases, that make air dirty

altitudes: distance above ground

atmosphere: the body of gases that surrounds Earth

climate: the usual weather in a region throughout the year

galaxies: large groups of stars

magnetic field: the area around Earth caused by attraction between the north and south poles

myths: old stories that often explain something in nature

particles: tiny bits of matter that can be solid, liquid, or gas

polar regions: the areas close to the north and south poles

radar: a system that uses radio waves to locate objects

solar system: the Sun and everything that orbits it

solar wind: the flow of particles away from the Sun

theory: an idea to test

Index